Arlindo Modesto Antunes
Ivano A. Devilla

Automation of a grain aeration system

Arlindo Modesto Antunes
Ivano A. Devilla

Automation of a grain aeration system

The construction of a prototype for grain aeration in metal silos

ScienciaScripts

Imprint

Any brand names and product names mentioned in this book are subject to trademark, brand or patent protection and are trademarks or registered trademarks of their respective holders. The use of brand names, product names, common names, trade names, product descriptions etc. even without a particular marking in this work is in no way to be construed to mean that such names may be regarded as unrestricted in respect of trademark and brand protection legislation and could thus be used by anyone.

Cover image: www.ingimage.com

This book is a translation from the original published under ISBN 978-3-330-74876-7.

Publisher:
Sciencia Scripts
is a trademark of
Dodo Books Indian Ocean Ltd. and OmniScriptum S.R.L publishing group

120 High Road, East Finchley, London, N2 9ED, United Kingdom
Str. Armeneasca 28/1, office 1, Chisinau MD-2012, Republic of Moldova, Europe
Printed at: see last page
ISBN: 978-620-7-89519-9

Copyright © Arlindo Modesto Antunes, Ivano A. Devilla
Copyright © 2024 Dodo Books Indian Ocean Ltd. and OmniScriptum S.R.L publishing group

Table of contents:

DEDICATE

My intellectual and professional formation could not have been achieved without the help of my loving and eternal parents Joel Antunes Pereira and Enedi Modesto da Silva Pereira, who, throughout my life, have given me, in addition to extensive affection and love, the knowledge of integrity, perseverance and always looking to God for the greatest strength for my development as a human being. For this reason, I would like to dedicate and acknowledge to you my immense gratitude and eternal love.

ACKNOWLEDGMENTS

First of all, to the God of Israel, author and creator of everything, for having supported and guided me through every moment of my life, giving me infinite victories.

To my family who have always encouraged me, supported me and dedicated their lives to my personal development. In particular, my parents, Enedi Modesto da Silva Pereira and Joel Antunes Pereira, my sister Dâmaris Modesto Antunes and my grandmother Ziza Modesto Rodrigues. Everyone has always been my secure foundation, even in the most difficult times.

I couldn't fail to mention my uncle Valdimar Adolfo da Silva and my aunt Ruth Andrade, who were undoubtedly responsible for giving me the strength and conditions to be able to achieve this milestone in my career.

To the friends who were essential for the elaboration and development of this project: Msc. Antônio Cruvinel, Msc. Guiuliano, Msc. Beethoven Gabriel, Msc. Danilo Gomes de Oliveira, Msc. Elson Antunes, Igor, Felipe de Oliveira Dourado, Marcos Paulo, Msc. Zeuxis Evangelista.

In particular, to Professor Elton Fialho dos Reis, whose extraordinary support enabled this project to be fully developed.

To my family members who welcomed me with open arms, dedicated part of their time and money to supporting me, sparing no effort, providing me with total support to complete this master's degree. They are: Ivo Pessoa, Cremilda Modesto, Lídia Regina, Leonardo Costa, Júlio César, Lourival Adolfo, Elisamar Gomes, Suzanna Marielle, Ellís Lorena, Wesley Gomes, Elcia Modesto, Paulo Pinto, Laudimar Modesto, Isneia, Misael Modesto.

To the staff of the Laboratory of Drying and Storage of Vegetable Products, Mr. Waldeir, for his help and dedication in making this work.

And to the person of Professor Dr. Ivano Alessandro Devilla who encouraged me and collaborated with his extraordinary knowledge to prepare this paper.

To everyone, my eternal gratitude.
Those who trust in the LORD will be like the mountain of Zion, which does not shake
but stands forever.
Psalms 125:1

SUMMARY

Grains must be preserved as much as possible during storage, because chemical, biochemical, physical and microbiological changes can occur. Grain storage consists of storing this material for use after a certain period of time. The aim of this work was to develop a prototype silo to evaluate different strategies for automated aeration of stored corn grain. A prototype metal silo with a cylindrical shape (0.9 m in diameter and 3 m high) with a perforated bottom was built and adapted to an aeration system, where the air flow was determined and maintained at $0.1 \text{ m}^3.\text{min}^{-1}.\text{t}^{-1}$. Sensors were used to collect data on the different aeration strategies, positioned in the center of the silo prototype, at 0.5 m intervals from the plenum. The sensors were connected to an Arduino data acquisition board in order to record the temperature of the grain mass. After installing the sensors in the silo, it was loaded and the surface of the grain mass was leveled. Control strategies were implemented: continuous aeration; dry bulb temperature control (Tbs< 22°C); timer control and temperature control via the temperature difference between the grain and the ambient air (TDIF=3°C). The operational results obtained with the computer program developed proved to be efficient for controlling the grain aeration processes stored in the prototype. It can be concluded that the system for automated aeration of stored grains was effective in storing grains and that the control strategy, aeration of grains via the temperature difference between the grains and the environment and aeration at night were the best strategies for controlling aeration in the region of Anápolis - GO.

Keywords: Storage, grain cooling, prototype, arduino.

Chapter 1

1 INTRODUCTION

Grain storage dates back centuries, with biblical references, when Joseph advised Pharaoh to store all the grain produced in Egypt, in times of plenty, to ensure the supply of these foods to Egyptian cities in times of extreme drought. Grain storage then consists of storing this product for consumption after a certain time (ARAÚJO et al., 2012).

It is well known that after being harvested, grains go through a series of processes until they reach the final consumer. Some of these operations are: reception, cleaning, drying and storage, handling, among others. According to Baal (2014), these operations are called grain pre-processing, and the places where they are carried out are called Grain Processing Units. Generally, grain processing sites are also made up of storage structures, so these sites can be called Grain Processing and Storage Units.

Grains must be preserved as much as possible during storage, due to the occurrence of chemical, biochemical, physical and microbiological changes. The speed and intensity of these processes depend on the intrinsic quality of the grains, the pre-processing operations, the storage system used and the environmental factors prevailing during the storage period (ALENCAR, 2006).

The main factor influencing the entire storage process is the grain's water content. Deterioration reactions occur at high humidity levels, especially enzymatic hydrolysis and lipid oxidation, but these reactions occur less frequently at low water contents. For this reason, the water content of grains must be controlled during storage (FURQUIM et al., 2014).

Moisture migration varies according to the season. During the winter and fall periods, grains located close to the silo walls and at the top of the grain mass are cooled more easily than those located at the bottom of the silo. After a while, due to the temperature gradient in the grain mass, convective currents are generated. In other words, the cold and dense intergranular air located near the silo walls is pulled downwards, flowing through the center of the silo and pushing up the warm and less dense air initially located in this region. In spring and summer, the temperatures of the grains near the silo walls increase and the grains located in the center of the silo remain cool. During these periods, the convective currents change direction. The cold, denser air located in the center of the silo flows downwards, resulting in a movement of convective currents from the center of the silo towards its sides (MUIR and JAYAS, 2003; VASCONCELLOS, 2012).

Currently, aeration is the most widespread control method used to preserve stored grains. Aeration consists of the forced passage of ambient air through the grain mass in such a way as to modify the intergranular microclimate, creating unfavorable conditions for the development of organisms that influence the preservation of grain quality. However, aeration is an operation that, if not carried out properly, can lead to loss due to heating, fermentation and excessive loss of water content, and is highly dependent on local climatic conditions. Therefore, the efficiency of an aeration system is centered on achieving a uniform air flow in all regions of the silo. Another aim of aeration is to prevent the migration and condensation of moisture that occurs whenever there is heating at any point in the grain mass (LUIZ, 2012).

According to Lopes (2006), aeration management should be implemented based on comprehensive studies of the devices that will be used. Aeration management is directly related to a control strategy, as it consists of activating fans based on the temperature and humidity conditions of the grains and the air.

Given the need to control and optimize the aeration process of stored grains, activities are being developed to automate the data acquisition system and consequently reduce the cost of the grain storage process (KALIYAN et al., 2007; LAWRENCE and MAIER, 2011; DEVILLA et al., 2012).

In view of the above, the aim was to:

1. Develop a computer program on the PHP platform to control aeration processes;
2. Building a prototype for aerating stored grain;
3. Evaluating strategies for controlling aeration of stored corn grains;
4. To collect data on the cooling and water content of corn grains stored in a metal silo, using different aeration strategies.

Chapter 2

2. LITERATURE REVIEW
2.1. General Features

Brazil, a country of continental dimensions with more than 376 million hectares, despite having great agricultural potential, is a country with a diverse climate, allowing for the production of various types of food. Because of this, grain production for the 2015/2016 harvest, according to the fifth report by the National Supply Company, is estimated at 210.3 million tons, an increase of 1.3% over the previous harvest (CONAB, 2016).

Corn (*Zeamays*) is one of the most widely grown cereals in Brazil and has a high nutritional quality. It is used as human food or animal feed. In Brazil, the state of Rio Grande do Sul stands out as the main national producer of first crop corn, with productivity reaching 40 tons per hectare (CONAB, 2015).

2.2. Grain Storage

Since the dawn of human history, food production has always been the most important thing in any society. The production, transportation, processing, storage, marketing and consumption of food is a chain of vital activities for people, families and nations, which is why agricultural storage is one of the oldest and most important activities for rural producers (WEBER, 2005).

The introduction of bulk grain handling and storage is a universal trend. In developed countries, bulk handling is widespread and integrated from harvest onwards. As farmers improve their level of technology, they tend to handle their production in bulk, as is the case in some regions in the south and southeast of the country (WEBER, 2005).

With regard to the planting area and production for the 2015/2016 harvest, the former is estimated at between 57.89 and 58.92 million hectares, representing an increase of up to 1.6% compared to the 2014/2015 harvest, which totaled 58 million hectares. As for production, the estimate is also for growth, and could reach values of 208.61 to 212.93 million tons, this result represents an increase of 64.7 and 4384.4 thousand tons over the crop harvested in 2014/2015 (CONAB, 2015). These characteristics show the full development of Brazilian agriculture.

Given the need to store grains harvested in the field, the aim of storage is to preserve the characteristics that the grains have after harvest. It is known that the vitality of the grains can be preserved and the milling quality and nutritional properties of the food can be maintained (BROOKER et al., 1992). However, it is scientific knowledge that grain quality cannot be improved during storage. Grains harvested inappropriately will be of poor quality no matter how they are stored, but good conditions during this period are essential to preserve the initial quality. Seed moisture and air temperature are variables that determine changes in quality during storage (BROOKER et al., 1982).

2.2.1. Storage in Metal Silos

Although there are biblical accounts of grain storage from the time of Joseph of Egypt, it is clear to see that storage systems have evolved a lot with technology, and it is possible to identify changes in their capacity and shape (KNOB, 2010).

Metal silos are found in all agricultural regions in Brazil and abroad (SANTOS and CENTENARO, 2014). To this end, grain storage units are characterized by watertight, airtight or semi-airtight cells or compartments, made of metal sheets screwed together to form a ring that overlaps to form a cylinder, also called a tube. They have the advantage of being quick to assemble and are available on the market in different diameters and heights.

Figure 1- Metal silos. Source: Santos and Centenaro (2014).

During the summer, solar heat can increase the temperature of grain stored in metal silos. The incidence of the sun's rays on the roof and walls of the structures is not directly responsible for the temperature changes in the mass of the grains, as they have low thermal conductivity. However, the reflective surfaces of the structures on the outside can improve the thermal conditions of the storage unit (SANTOS and CENTENARO, 2014).

2.3. Aeration

According to studies, aeration is the adequate movement of ambient air through the grain mass to improve storage conditions (SAUER, 1992; JAYAS et al., 1995). Aeration is the forced passage of air through the grain mass in order to preserve quality or solve storage problems (NAVARRO and CALDERON, 1982; HAGSTRUM et al., 1999; SILVA et al., 2000). Lasseran (1981) defines aeration or ventilation as the forced circulation of ambient air through the grain mass.

According to Júnior (2011), the purpose of aeration is to maintain the stored grains, without affecting the quality and quantity of the stored mass, through a mechanical ventilation system. Among its many uses, the main ones are:

2.3.1. reducing grain temperature and humidity;

2.3.2. insect and fungus control;

2.3.3. avoid convection air currents;

2.3.4. avoid mechanical damage;

2.3.5. preserve the physical and chemical qualities of grains.

Studies have shown two distinct characteristics of the grain aeration period. Firstly, it is known that over a shorter aeration period, the temperature of the grains seeks to match that of the environment, and may even decrease in relation to the environment, which can be called cooling. Secondly, after a long period of aeration, the effect may be that the grains dry out. These effects can be explained by the laws of equilibrium between the air and the grain (USATIKOV et al., 2003).

Because of these detailed characteristics of the process, inefficient management of the aeration process can result in product deterioration, consequently losses from a nutritional point of view and a reduction in commercial value (MAGAN and ALDRED, 2007; REED et al., 2000; KHATCHATOURIAN et al., 2016).

The minimum specific air flow required depends on the type of grain being stored, the thickness of the grain mass, the type of installation and the number of storage structures in the system. Generally, in cold climates, as shown in Table 1, with horizontal structures, the specific flow rates for aeration systems vary between 0.05 and 0.1 m^3 min^{-1} t^{-1} . In vertical structures, these flow rates vary between 0.03 and 0.05 m^3 min^{-1} t^{-1} . In hot regions the specific flow rates vary between 0.1 and 0.2 m^3 min^{-1} t^{-1} for horizontal structures that store dry grains and between 0.03 and 0.1 m^3 min^{-1} t^{-1} for vertical structures (SILVA et. al., 2000).

Table 1 - Air flow recommendations for aerating grain in silos.

Type of unit/purpose_	Air flow (m^3 min^{-1} t^{-1} of grain)	
	Cold region	Hot region
Horizontal / dry grain	0,05 a 0,10	0,10 a 0,20
Vertical / dry grain	0,02 a 0,05	0,03 a 0,10
Lung / wet grains	0,30 a 0,6	0,30 a 0,60
Drought-aeration	0,50 a 1,00	0,50 a 1,00

Source: Silva (2008).

2.3.6. Cooling the Grain Mass

In cold and temperate regions, cooling the grain is the main objective of aeration. With relative humidity within the ideal parameters, ambient air is introduced into the storage environment, gradually lowering the temperature of the grain mass and creating a microclimate unsuitable for insect proliferation. However, it is difficult to achieve low temperatures and water contents using aeration in countries with tropical and subtropical climates, such as Brazil. In these regions, it is recommended that grains are stored dry (between 11 and 13% b.u.) and that aeration is used with the main objective of maintaining homogeneous temperatures in the grain mass (LACERDA FILHO and AFONSO, 1992).

Considering tropical and subtropical climates, aeration has proven to be an efficient process, especially in regions with cold nights and winters. Research carried out in Canada and the northeastern United States has shown that grain masses cooled during the winter remain at a low temperature during the spring and summer (CASADA et al., 2002 and JAYAS and WHITE, 2003).

2.3.7. Temperature homogenization

The use of aeration in order to maintain a low temperature gradient in the storage environment makes it possible to prevent the migration of moisture, hot spots and water condensation in the stored grains (BARRETO, 2013).

The profile of moisture migration varies according to the time of year; in winter and autumn, grains located close to the silo walls and at the top of the grain mass are cooled more easily than those located at the bottom and in the center of the silo. After a few days, due to the temperature gradient in the grain mass, lateral convective currents are generated (USATIKOV et al., 2003). In other words, the cold, dense intergranular air located near the silo walls is pulled downwards, flowing through the center of the silo and pulling up the warm, less dense air initially located in this region.

In spring and summer, the temperatures of the grains near the silo walls increase and the grains located in the center of the silo remain cool. During these periods, the convective currents change direction. The cooler, denser air located in the center of the silo flows downwards, resulting in a movement of convective currents from the center of the silo towards its sides (USATIKOV et al., 2003).

2.3.8. Grain aeration strategies

Aeration strategies should be chosen based on the characteristics of the stored grain, the water content of the grain, the storage location, the temperature of the product and the temperature of the environment and the classification of aeration types. Some existing classifications of aeration, according to Weber (2005), are presented below.

Maintenance aeration: in this case, the grains are stored clean, dry and cold. The aim in this case is to maintain the grains by neutralizing the spontaneous heating of the grains. This type should be used when the temperature difference between the internal and external environment reaches 6°C in cold climate regions. In regions with a hot climate, the system should be activated when it is possible to reduce the temperature of the grains by 2 to 3 °C.

Corrective aeration: whenever a high temperature is observed in any region of the grain mass, due to a high concentration of impurities, insects or moisture infiltration, corrective aeration should be initiated.

High temperatures, above those recommended for preservation, are common. This is caused by the following:

- occurs when the grains, after passing through the dryer where they dry and cool, even "cold", the temperature will be above the ambient temperature, which represents a high temperature, especially on hot days.
- This refers to the drying strategy, in which the dryer is used using a continuous or intermittent system, with full-body drying, without cooling in the dryer. In this case, the grains come out hot and are cooled in the silo.

With this in mind, studies show the need to study the diversity of strategies for controlling the aeration of stored grain, comparing forms of aeration, economic viability and the performance of different types of aeration, in order to determine the best form of aeration (NASCIMENTO and QUEIROZ, 2011; RANALLI et al., 2002).

Nascimento and Queiroz (2011) evaluated and compared the performance of three different aeration strategies for stored corn: continuous aeration, night aeration and aeration under conditions of hygroscopic equilibrium between the grains and the ambient air. They observed that continuous aeration was efficient on moist grains and none of the aerations maintained a homogeneous temperature inside the grains during storage.

A study carried out with the aim of evaluating two strategies for controlling and managing aeration on the quality of corn grain in metal silos showed that dry bulb control combined with relative humidity were not enough to have a cooling effect on the grain mass (RIGO et al., 2012).

2.4. Computerized Aeration

There are currently manually controlled aeration systems where, simply by monitoring the temperature and humidity of the stored grain, the operator activates the fans, leaving them running for as long as necessary to reach the desired temperature and humidity, and then switches them off. Much more sophisticated automatic controllers are also used for the same purpose, but on a smaller scale, given the lack of knowledge of this technology and its higher cost (LOPES et al., 2010).

Automatic aeration systems can work with the help of sensors, thermostats and humidistats, and can be coupled to a control panel and passed through a microprocessor, associated with a microcomputer and auxiliary electronic devices. Depending on the temperature and humidity of the grain mass, as desired and elaborated in the automation system, the system is adjusted to turn the fans on or off, according to the pre-established conditions. With aeration systems, using a microcomputer with a video terminal and printer available, data can be recorded at any time, as described by (EVANS et al., 2013; PUZZI, 2003).

Aeration management must be implemented based on comprehensive studies of the devices that will be used, the storage structure, the local climatic conditions and the characteristics of the aeration system. Essentially, aeration management is directly related to a control strategy, which consists of activating fans based on the temperature and humidity conditions of the grains and the air (LOPES, 2006).

Lopes et al. (2008) developed control strategies for stored grain aeration systems. The strategies were implemented in software called AERO, which can be configured to maintain optimal grain storage with minimal energy expenditure or to prioritize greater cooling of the grain mass.

Lopes et al. (2010) carried out an ongoing study with the aim of comparing, by means of simulations, the effects on the storage environment of the control strategies used by AERO and by controllers based on timers, thermostats combined with humidistats. The results obtained from the simulations indicate that AERO and the thermostat- and humidistat-based controllers are efficient at controlling the grain storage environment.

Other studies with computerized aeration, using controllers such as OPI 2000, Air Master and Air Control Draw, are managed by computer programs that activate the aeration system according to data measured by a weather station and thermocouples or thermistors installed in the grain mass. In these controllers, the strategies are based mainly on the equilibrium water content equation of the grains and on temperature limits (KALIYAN et al., 2007; LOPES, 2010).

Software capable of simulating the aeration process offers other advantages, such as the possibility of using it to investigate the effects of the various parameters that affect the aeration process, taking into account the different sizes and types of silo, the type of wall material, the type of grain stored and the location of the experiment, which can be in different regions (LOPES et al., 2008).

However, computerized programs and processes are not used as much in storage, because producers are unable to adapt to the complexity of using the equipment used to manage aeration. They are therefore forced to abandon the system due to its "uselessness" (LOPES et. al., 2008).

2.5. Arduino microcontrollers

Since the Arduino Project began in 2005, more than 500,000 Arduino boards have been sold worldwide. Unofficial clone boards undoubtedly outnumber official boards, and it is likely that over a million Arduino boards and their variants have been sold. Its popularity continues to grow and more and more people are realizing the sensational potential of this incredible open source project and its ability to create interesting projects quickly and easily, with a relatively low learning curve (CAVALCANTE et al., 2011).

The use of Arduino as a platform for controlling the aeration process of stored grain is already a reality today, making it a viable way of reducing costs in the storage process (WHITE, 2013).

The use of this control platform requires programming for the aeration environment and promotes communication between the office and the silo, acquiring temperature and humidity information from the sensors in the silo (ZHANGXIAO RU, 2013).

Chapter 3

3. MATERIAL AND METHODS

3.1. Characterization and Location of the Study Area

This work was carried out in the Laboratory for Drying and Storing Plant Products at the Campus of Exact and Technological Sciences Henrique Santilo, of the State University of Goiás in Anápolis-GO. The laboratory is located on the Brazilian Central Plateau, with latitude 16°19' 36" S and longitude 48° 57' 10" O and an altitude of 1,017 m, with a tropical climate and temperatures ranging from 18° to 36°C (IMB, 2014).

3.2. Grain characteristics

The experiments used maize grains (*Zeamays* L.), with an initial water content of 12.6% (b.u.), purchased by the State University of Goiás from local retailers. The corn grains were treated with phosphine before starting the experiment to control pests.

3.3. Characterization of the equipment that makes up the aeration system

3.3.1. Silo

A cylindrical metal silo was used (0.9 m in diameter and 3 m high), made of smooth material, with the capacity to store 1500 kg of corn grain, with a specific weight of 750 kg m^{-3} .

The plenum was built using building materials such as bricks, cement and sand. Its construction was designed to allow air to enter the silo, and its dimensions were (1.2 x 1.2 x 0.30 m), Figure 2.

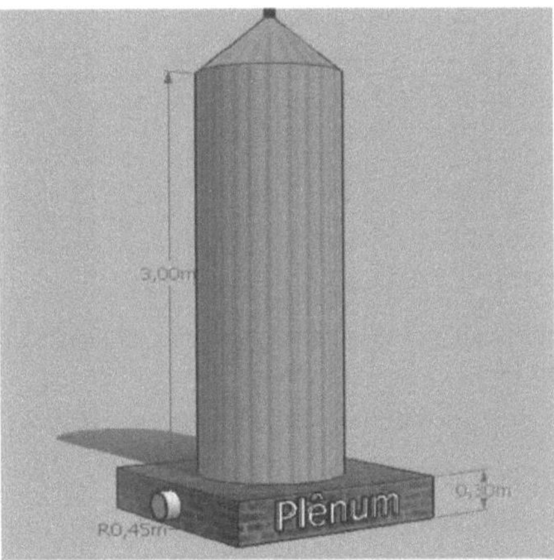

Figure 2 Schematic of the grain storage prototype used to evaluate the different aeration strategies. Source: author.

The thermal insulation (edges of the silo) was built with a layer of glass wool, 0.06 m thick, in order to insulate it. At the bottom of the silo, sheets of perforated auger with 0.006 m diameter sieves were installed to promote the passage of air through the grain mass, as shown in Figure 3.

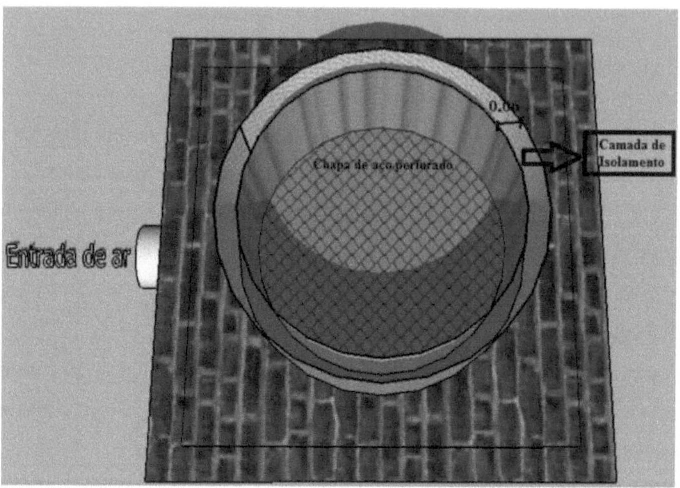

Figure 3 Top view of the corn grain storage prototype. Source: author.

3.3.2. Fan.

An aeration system consisting of a radial-bladed centrifugal fan (Figure 4) with a 1 hp three-phase motor, made of sheet metal and designed to supply an air flow rate of 6 m^3 min^{-1} was fitted to the silo plenum. A device was fitted to restrict the air inlet, thus achieving an air flow rate of 0.10 m^3 min^{-1} t.$^{-1}$

Figure 4 Diagram of the centrifugal fan with radial blades used to blow air into the silo.

4.3.2.1. Determination of air flow

To determine the air flow, the methodology described by Delmée (1982) was used, in which a Pitot tube was used to determine the air flow in pipes. To do this, a PVC pipe with a diameter of 0.2 m and a length of 2.0 m was attached to the fan (Figure 5). The pitot tube was inserted into the pipe through a hole in the pipe, located 1.7 m from the fan. An air flow homogenizer was placed in the pipe, made of square-sectioned cells, 0.7 m from the fan's air outlet.

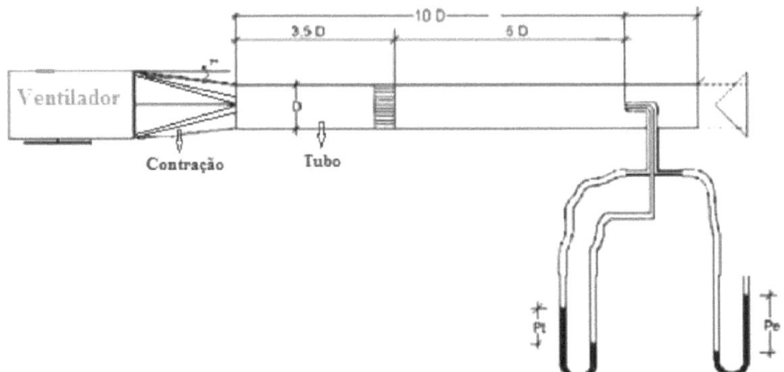

Figure 5 Pitot tube assembly used in the process of determining fan air flow. Source: Macintyre (1990).

Air flow was determined by the difference between total pressure (Pt) and static pressure (Pe), using a manometer with a scale of 700 Pa and an accuracy of 0.09 Pa, installed in the Pitot tube.

The evaluation of the air coming from the fan was determined using Bernoulli's equations (Equation 1 - conservation of energy). It is therefore necessary to estimate the average velocity from the variations in pressure exerted by the fluid in closed ducts (DELMÉE, 1982; MACINTYRE, 1990).

$$\frac{p1}{pe} + \frac{v^2}{2g} = \frac{p2}{pe} + \frac{v^2}{2g} = constante \qquad (1)$$

in which:

P= total pressure measurement, m.c.a;

v= average fluid velocity, m s^{-1} ;

Pe= Specific weight of the fluid, N/m^3 ;

g= acceleration due to gravity, m s^{-2} .

and consequently the air flow velocity was calculated from equation 2.

$$v = \sqrt{2.g.h} \qquad (2)$$

in which:
g = acceleration due to gravity, m s^{-2} ;

h = manometric height, m.c.a.

Finally, the fan's air flow rate was determined using Equation 3 (MACINTYRE, 1990).

$$Q = A.v \qquad (3)$$

in which:

Q= fan air flow, m^3 s^{-1} .

A= cross-sectional area of the pipe, m2;

V= air flow velocity, m s^{-1} .

3.3.3. Heating System

In order to achieve a grain temperature of around 30°C after the drying process, a device was developed to heat the stored grain (LAWRENCE and MAIER, 2011).

The device consisted of a metal box with dimensions of 0.50 x 0.30 x 0.35 m, Figure 6. Two finned resistors (type U) with dimensions of 0.45 x 0.06 m and a power of 1 kW were installed in the metal box, Figure 7.

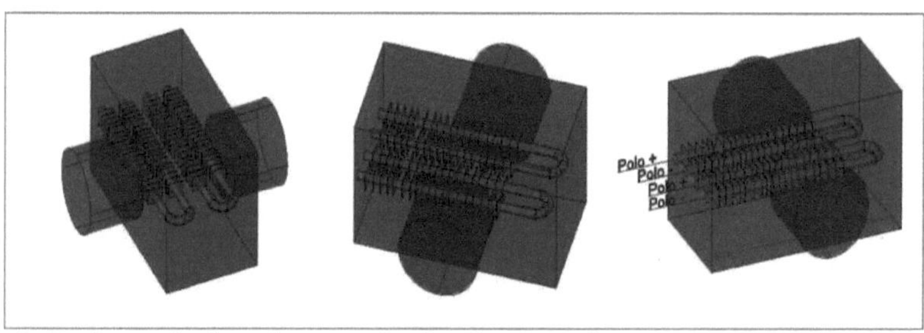

Figure 6 - Device used to heat the mass of grains. Source: author

The heating system was coupled to the fan and to a 0.10 m diameter PVC pipe that carried the air to the silo plenum. Before all the aeration tests, the device was attached to the silo and the fan to heat the grain mass.

Inside the metal box, the resistors were insulated with rubber and the electrical connection to the power source was made using cables 0.002 m in diameter and 5 m long. To monitor the heating process, a Cycloar digital thermo-hygrometer with a resolution of ± 0.1°C was installed in the PVC pipe before the plenum.

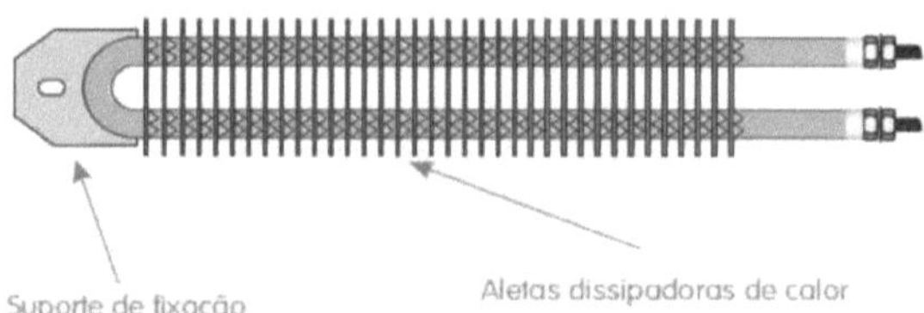

Figure 7- U-type finned resistor used to heat the mass of corn grains in the silo. Source: author himself.

3.3.4. Recirculation

In order to prevent losses in the water content of the grains to the outside environment, a closed system was developed, with the aim of recirculating the aeration air and the consequent gradual increase in temperature (LAWRENCE and MAIER, 2011).

PVC pipes were connected to the silo (Figure 8). The air blown in by the fan passes through the heating device and then enters the plenum at the bottom of the silo. A digital thermo-hygrometer was placed before the plenum to monitor the air temperature.

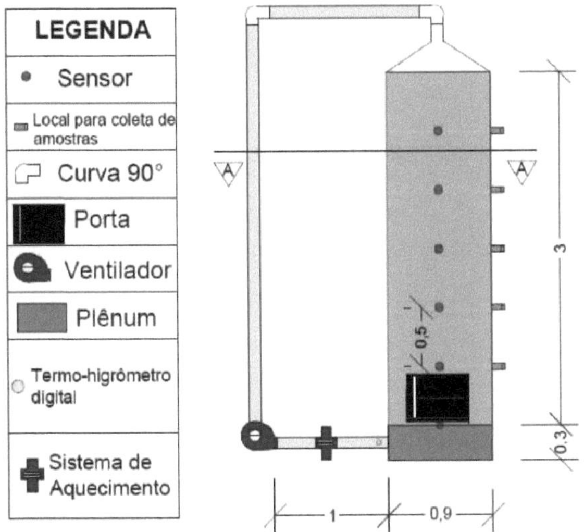

Figure 8 - Sketch of the silo with air circulation system used to evaluate the strategies. Source: author himself.

3.4 Control of the aeration system

The control of the aeragaofoiconceived by means of:

- Sensors (temperature and relative humidity);

- Arduino Nano microcontroller;

- Relay board;

- Contactor and Thermal Relay;

- Microcomputer; and

- Fan;

3.4.1. Sensors

Six SHT 75 sensors, which indicate the temperature and humidity of the interstitial air with an accuracy of 0.001°C, were installed in the central part of the grain mass, Figure 9. The sensors were distributed in the grain mass, spaced 0.5 m apart, Figure 8.

Figure 9- Sensor used to monitor temperature and relative humidity in the center of the silo, model SHT 75. Source: author.

Environmental parameters were also recorded using the SHT 75 sensor, positioned outside the grain mass. Data was acquired every 1 minute and stored on a microcomputer, as shown in Figure 10.

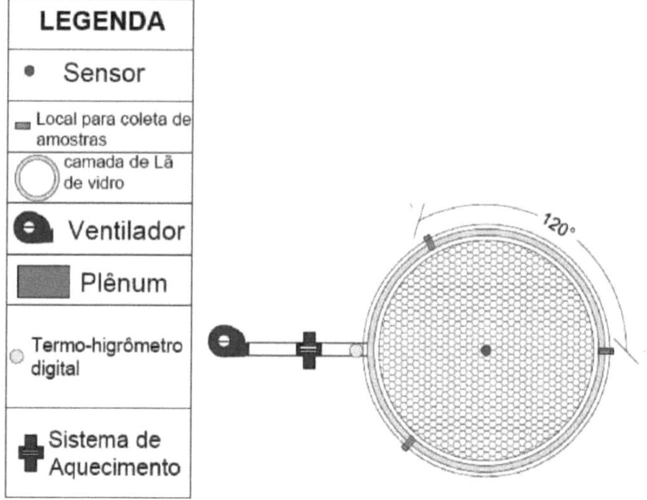

Figure 10- Cross-section of the silo, showing the location of the sensors inside the grain mass. Source: author himself.

The silo was filled by vertical discharge. A device was therefore created to avoid damaging the sensors, which were placed in the center of the silo. This device was made up of a quarter-inch (%") tube and was approximately

0.1m long, the entire pipe was perforated. The sensor was then stored inside the pipe and capped, Figure 11.

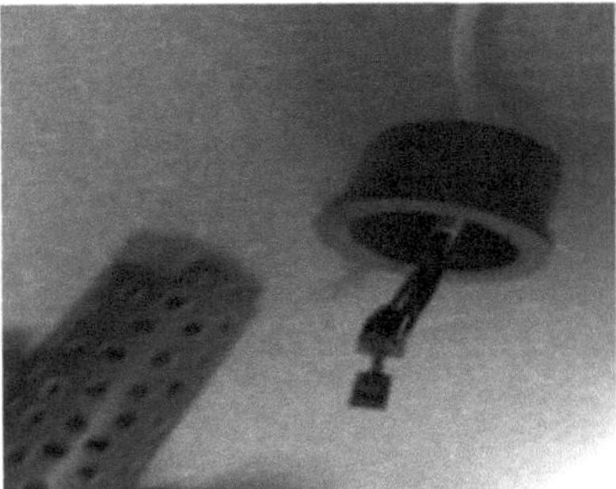

Figure 11 - Protection device installed next to the temperature and relative humidity sensor. Source: author himself.

3.4.2. Arduino Nano microcontroller

The sensors were positioned in the silo, above 0.05 m from the plenum, inside the grain mass, totaling six sensors, in order to collect data from the aeration environment. All the sensors were connected to a four-way sleeve cable, Figure 12. The cable was connected to an Arduino Nano micro controller in order to record the temperature of the grain mass during the aeration process, Figure 13.

The temperature of the grain mass was recorded using an Arduino Nano microcontroller, an open source prototyping platform based on user-friendly hardware and software, which was used to record the temperature of the grain mass (ZULKIFLI et. al., 2012).

The microcontroller in question is based on an Atmel AVR reduced instruction set computer (RISC) 8-bit microprocessor, which runs at clock speeds of over 16 MHz and has 128 KB of flash memory (EVANS et al., 2013).

Figure 12- Four-way sleeve cable used to connect the Arduino Nano microcontroller. Source: author himself.

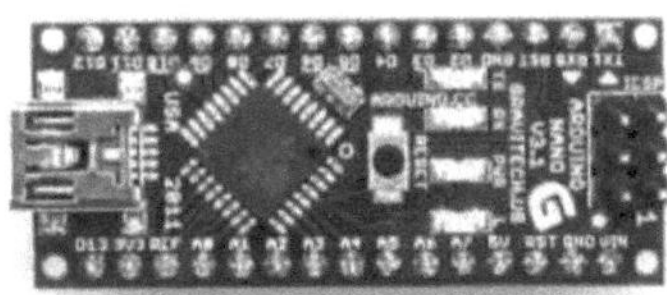

Figure 13- Arduino Nano 3.x used in this work. Source: author.

3.4.3. Relay board

To turn the fan on and off, a relay bank (*single double throw SPDT*) was used, which supports a voltage of 380 volts, 10 A of current. For the process, a relay was connected to one of the phases of the electricity network, which was responsible for triggering the contactor coil responsible for switching the fan on and off (Figures 14 and 15).

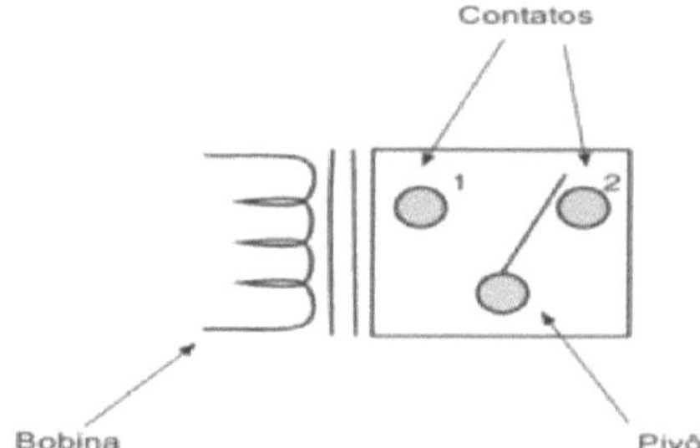

Figure 14- Elements of the SPDT relay used to switch the aeration system fan on and off. Source: Evans et al. (2011).

Figure 15 - Relay board used to switch the grain aeration system fan on and off. Source: author himself.

3.4.4. Contactor and Thermal Relay

In the event of an electrical overload, in order to preventively interrupt the system, a Siemens contactor (3RT1015), which supports 9 A of current and 380V/60 Hz coil voltage, was connected to the Siemens thermal relay (3RU1116), working with the motor's nominal current of 8 A. This equipment is used for preventive purposes, since it performs the function of remote reset and interrupts the system in the event of an electrical overload. For a better demonstration of the connection process, the three-phase connection diagram is shown in Figure 16.

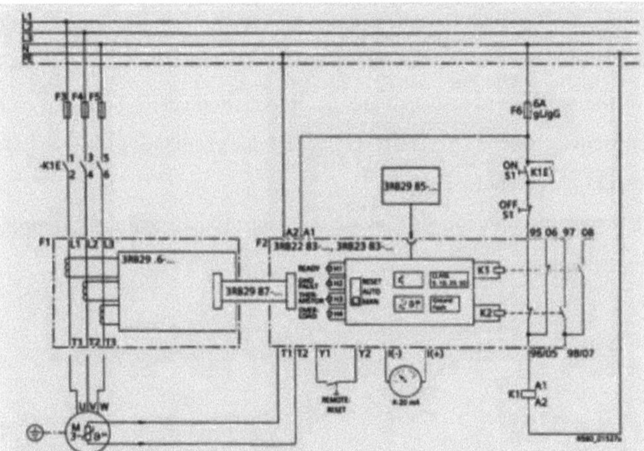

Figure 16 - Three-phase wiring diagram for the 3RT1015 contactor used to interrupt the system in the event of an electrical overload. Source: Siemens Catalog.

3.4.5. Microcomputer

To constantly monitor the aeration process and share a screen for remote access to the system, TeamViewer software was installed on a Sony Xperia T2 cell phone and on the microcomputer. This helped with the "on and off" process of the aeration system. However, the Arduino nano micro controller was controlled by the microcomputer.

The microcomputer used for data processing has the following configuration: Intel Dual Core microprocessor, processor speed 2.30 GHz, HD 1 Terabyte and RAM 4GB DDRII 800 MHz, running Windows XP professional operating system, 32 bits, Figure 17.

Figure 17 - Microcomputer used to acquire and process data from the Arduino system. Source: author.

3.4.6. Fan drive

The main structure for driving the fan is based on the functionalities of the Arduino Nano board and its Arduino 1.0.1 system, installed on the microcomputer, which control the temperature and relative humidity sensors.

As soon as the sensors are activated and the information comes back, which happens every minute, the system activates the fan via the relay board, which responds to the criteria established in topic 3.6 Validation of the computer program.

Figure 18 shows the operating stages of the control and data acquisition system used to evaluate the different strategies of the grain aeration process.

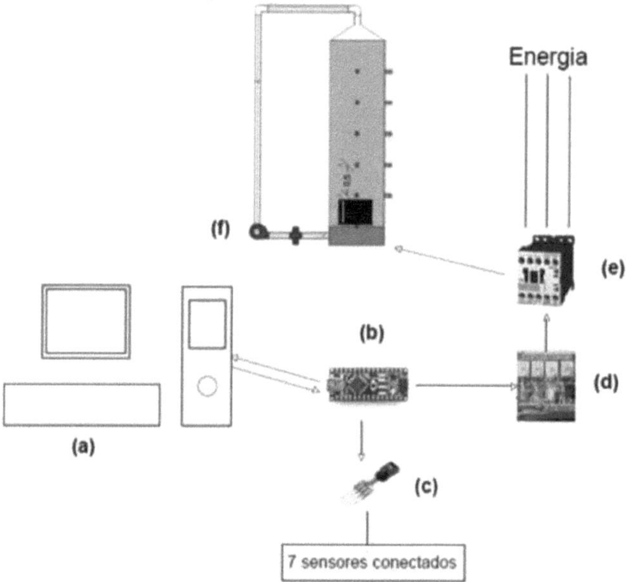

Figure 18 - Schematic of the airflow control and data acquisition system, (a) microcomputer, (b) Arduino nano, (c) sensors, (d) relay board, (e) contactor, (f) fan motor. Source: author himself.

3.5. Data Acquisition System

Data analysis and the decision to turn the fan on or not comes from the attributes defined in the algorithm developed on the Arduino 1.0.1 platform (Figure 19), the source code of which is shown in Appendix A.

The algorithm was developed using standard libraries provided by the Arduino software (MCROBERTS, 2011).

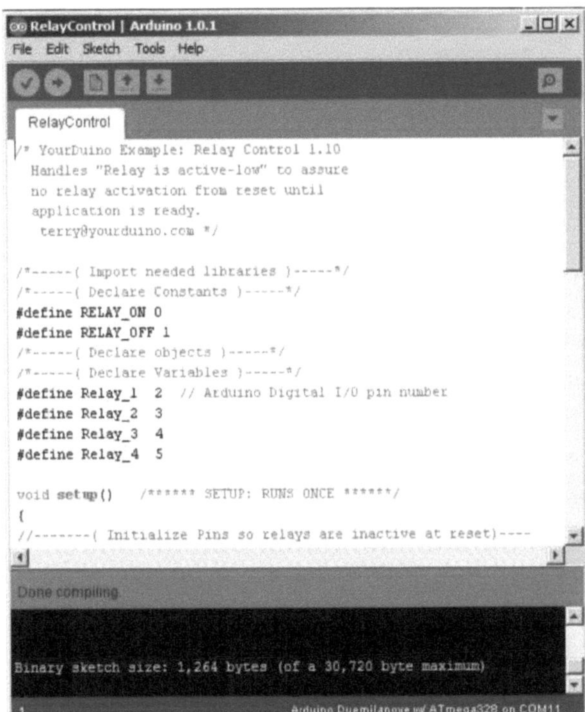

Figure 19 - Initializing the Arduino software. Source: author.

All the decision making for operating the fan was developed in this algorithm, using as parameters the data read by the temperature and relative humidity sensors located inside and outside the grain mass.

In order for the grain aeration system to work fully, so that data could be stored minute by minute, it was necessary to develop another system based on the PHP programming language. The program selected was Xampp, which uses the most popular language and is easily accessible to students (WALIA and GILL, 2014).

3.6. Computer Program Validation

After the entire system had been installed, the silo was loaded and the surface of the grain mass leveled.

The grain mass was heated, and the entire set (slipper hat, heating prototype and PVC pipes) was removed from the silo, enabling the computer program to be activated. Figure 20 shows the location on the grain mass of each sensor (1, 2, 3, 4, 5 and height) and the location for collecting grain samples to determine humidity. In order to control the dynamics of grain mass cooling, a sensor was installed 0.05 m from the base of the plenum.

Ambient sensor"

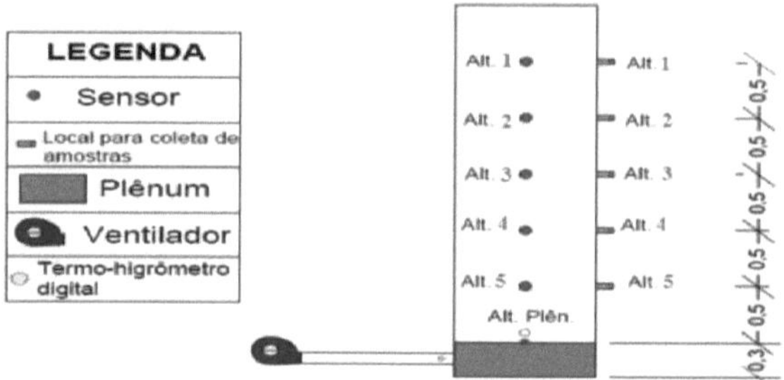

Figure 20- Sketch of the silo ready to start aeration, without the air heating device.

Evaluations of the software developed were carried out according to the following control strategies for the aeration system:

(a)   continuous aeration, fan on 24 hours;

(b)   ambient air temperature control (Tar< 22°C ±0.9);

(c)   timer control (fan on from 9pm to 10am); and

(d)   temperature control via the temperature difference between the grains and the ambient air (TDIF=3°C).

Continuous aeration was taken as a test for the other aeration processes, since it depends on the characteristics of the external environment. Therefore, if it is not carried out properly, it can cause the grains to overdry (MOREIRA, 1993).

The aim of ambient air temperature control is to allow the grain mass to cool down based on ambient temperatures below 22 °C. It is not widely used in tropical and subtropical regions, as it is a limited process that can lead to insect infestation (LOPES et al., 2010).

Taking this process into account, the systematization of control (b), by programmed logic, is shown in Figure 21.

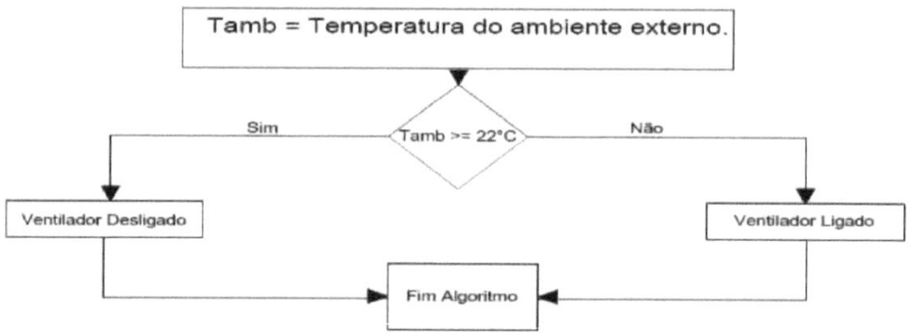

Figure 21 Diagram used by the room air temperature control strategy.

The timer control, item (c), allows the fan to be turned on during the night, when there are lower temperatures in the environment, allowing for uniform cooling of the grain mass. However, in some tropical and subtropical regions, there are high relative humidities at night, which can directly affect the water content of the grains (LAWRENCE and MAIER, 2011).

In order to simplify the timer control, its algorithm is detailed in Figure 22. In this algorithm, the fan is activated from 9pm until 10am. This algorithm was implemented on the Arduino 1.0.1 programming platform.

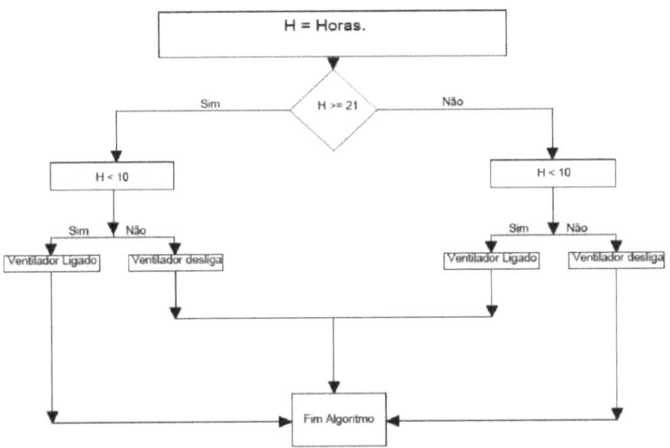

Figure 22 - Diagram used by the timer-based control strategy.

The fourth condition (d) allows the fan to be switched on when the temperature gradient exceeds 3°C. This condition guarantees homogeneous temperatures in the grain mass and is very important in tropical and subtropical regions. Therefore, in these regions, the main objective of aeration should be to operate the fan whenever the outside temperature is 3 °C to 5 °C lower than the grain mass, minimizing moisture migration. In practice, experts guarantee that temperature differences of up to 5 °C are acceptable (NAVARRO and NOYES, 2010; SILVA et al., 2000).

Figure 23 shows the simplified temperature control algorithm via the temperature difference between the grains and the ambient air (TDIF=3°C).

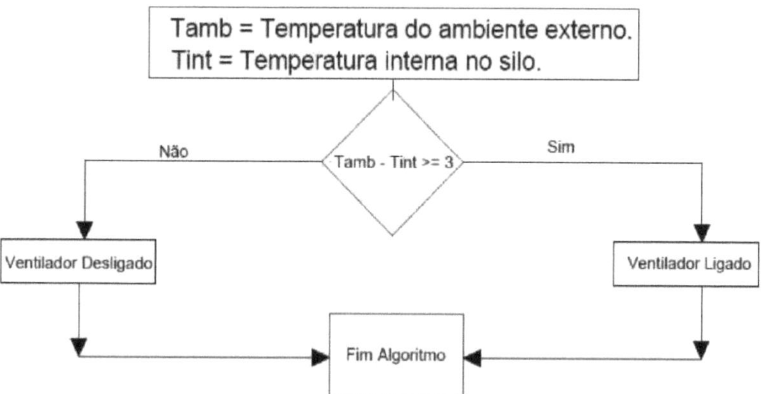

Figure 23 - Diagram used by the control strategy based on the temperature difference between the grains and the ambient air.

Since there is the possibility of requiring manual control, the possibility of "turning on" or "turning off" the fan was implemented, assigning the value "zero" (off) or "one" for (on), in any aeration strategy.

3.7 Grain water content

Grain samples were taken at the end of each aeration day (24 hours) to determine the final moisture profile. Moisture was determined using the oven method at 105 °C, with air circulation, for 24 hours, in three replications according to BRASIL (2009).

At the end of the aeration tests, the best strategy was selected according to the following characteristics: it cooled the grain; it conserved the water content without causing drying; it maintained cooling; it cooled in the shortest time.

Chapter 4

4. RESULTS AND DISCUSSION

4.1. Silo

The storage capacity of the silo was 1500 kg of maize grain (*Zeamays* L). Therefore, in this work, 30 50 kg bags of corn grain were stored and unloaded vertically into the silo. The top of the silo was accessed using a metal ladder, Figure 24. The data acquisition board, connections and microcomputer were installed inside a water tank to avoid damage.

Figure 24 - General view of the prototype silo where the different strategies for controlling the aeration of stored grain were evaluated.

4.2. Fan

The centrifugal fan with radial blades, made from metal, used a 1 hp three-phase motor connected by a 0.43 m diameter "v" belt. In order for air to pass through the PVC pipe, the fan's air outlet had to be reduced to 0.1 m (Figure 25).

Figure 25 - Fan (a), motor (b), belt (c) and reduction gear (d) used to blow air into the grain storage silo.

The air flow from the fan was controlled by means of a metal diaphragm positioned at the air suction inlet (Figure 26).

Figure 26 - Diaphragm attached to the fan to restrict the air flow into the silo. Source: author himself.

4.2.1 Determining Fan Flow

The results obtained from determining the fan's air flow using the pitot tube method are shown in Table 2. The average flow rate supplied by the fan was [0.00152± 0.00158] m^3 s^{-1} . At the center of the 0.2 m pipe, average values of [0.00198 ± 0.0001] itf s^{-1} were collected, as these are desired values (SILVA et al., 2000).

The pitot tube was placed at various depths inside the 0.2 m PVC pipe, as shown in Table 2 in the reading position. According to Macintyre (1990), the readings were spread out by 0.02 m.

Table 2 - Averages and standard deviation of velocity, air flow and manometric height in the different positions of the Pitot tube.

Readings	Reading position (m)	Manometric Height (Pa)	Speed m s^{-1}	Flow (LPM)	Vazaom^3 s-1
Top edge	0,02	0,980 ± 0,0	0,04 ±0,0	83,366±0,0	0,0014 ±0,0
	0,04	1,883 ±0,107	0,06±0,002	115,477 ± 3,314	0,0019 ±0,0001
	0,06	1,687 ±0,407	0,06±0,007	108,570±14,411	0,0018 ±0,0002
	0,08	1,765 ±0,0	0,06±0,0	111,847 ±0,0	0,0019 ±0,0001
Center	0,1	1,295 ±0,429	0,05±0,008	94,758±15,599	0,0016 ±0,0006
	0,12	0,981 ±0,0001	0,04±0,0001	83,366 ±0,0001	0,0014 ±0,0
	0,14	0,981 ±0,0	0,04±0,0	83,365 ±0,0001	0,0014 ±0,0
	0,16	0,981 ±0,0	0,04±0,0	83,365 ±0,0001	0,0014 ±0,0
	0,18	0,666 ±0,161	0,04±0,004	68,318±8,553	0,0011 ±0,0001
Lower edge	r0 ,2	0,883 ±0,098	0,04±0,002	78,989 ±4,400	0,0013 ±0,0001
	Overall Average	1,210±0,42	0,049±0,008	91,377±0,0002	0,00152±15,80

4.3. Grain Heating and Recirculation System.

The characteristics of the grain heating device are shown in Figures 27 and 28, where: the fan, the device and the PVC pipe (0.1 m) are connected to the plenum of the storage silo.

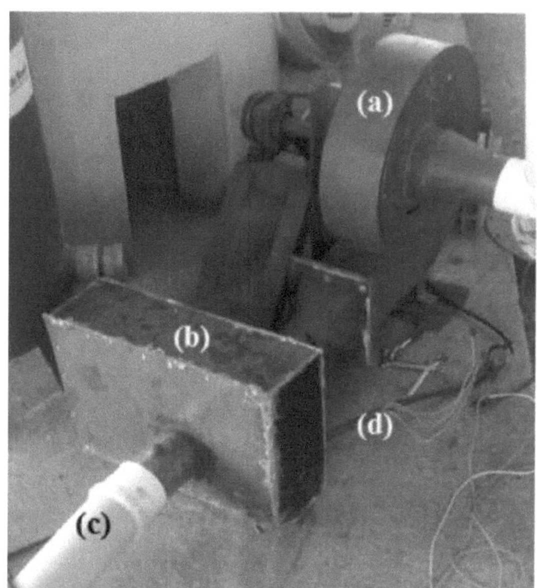

Figure 27 - Fan (a), air heating prototype (b), PVC pipe (c), power cables (d), used to heat the grain mass in the silo. Source: author himself.

The device, consisting of two 1 kW finned resistors (type U), worked properly when the resistors were arranged in series and connected to a 2.5 mm diameter cable (Figure 28). The heating system provided temperatures of up to 40 °C, which were monitored by the Cycloar digital thermo-hygrometer, placed just after the air outlet of the heating device.

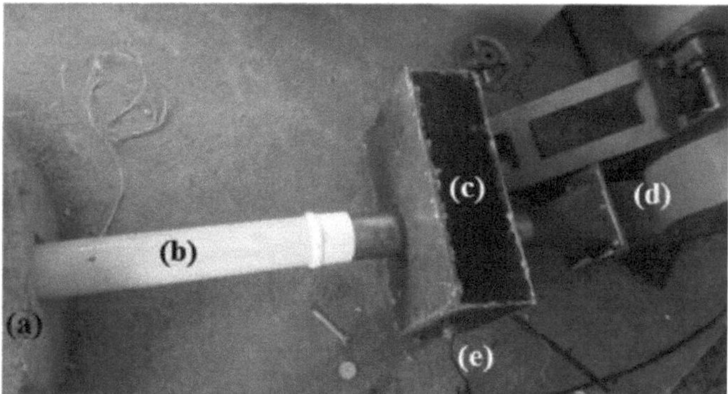

Figure 28 - Prototype for heating the aeration air (c), coupled to the PVC pipe (b), blowing air into the silo plenum (a), fan (d) and electrical cables (e). Source: author himself.

Table 3 shows the average time taken to heat and recirculate the air, in order to ascertain the time taken to heat the grain mass.

Table 3 - Mean initial temperature (TMI), mean final temperature (TMF) and heating time for each aeration strategy.

Strategy	TMI (°C)	TMF (°C)	Time (hours)

a	28,5±1,22	30,98±1,33	2,5
b	25,5±0,62	33,6±0,65	8,0
c	27,2±0,35	31,2±1,41	5,2
d	27,3±0,82	31,5±1,01	6,0

The process of heating the grain mass was faster in the continuous aeration strategy (a), as the initial temperature of the grain mass was 28.5 °C. The total elapsed time for heating and air recirculation was 2.5 hours to heat 2.4 °C.

The heating of the grain mass in strategy b, ambient air temperature control (Tar< 22°C ±0.9) was conducted for a prolonged period of time, due to the temperature of the outside environment being 23 °C. To complete the heating process, it took 8 hours to heat up 8 °C, given that the average initial temperature of the grain mass was [25.5±0.62] °C.

The uniform heating of the grain mass was made possible by the recirculation prototype developed for the experiment. Figure 29 shows that the PVC pipe was attached to the fan at the top of the silo. The heater was placed at the fan's air outlet, followed by a pipe connected to the plenum, making it possible to recirculate the air.

Figure 29 - Air recirculation system for heating grain, (a) fan, (b) silo, (c) PVC pipe and (d) heating prototype. Source: author himself.

4.4. Aeration System Controllers

It can be seen that the use of Arduino was adequate for the control process, since during the period of aeration of the grain mass, the microcontroller remained in perfect working order, controlling the cooling of the grain mass, according to the pre-established strategy. Other authors (CAVALCANTE et. al., 2011; EVANS et al., 2013; DILLY and MENDES, 2015) also cite Arduino as being functional for acquiring temperature and humidity data.

4.5. Computer Program Developed

The screens of the computer program developed are shown in Figures 30 and 31.

The initial screen, Figure 30, identifies the computer program, giving some basic information about it, such as its function, the name of the author and supervisor responsible for the project.

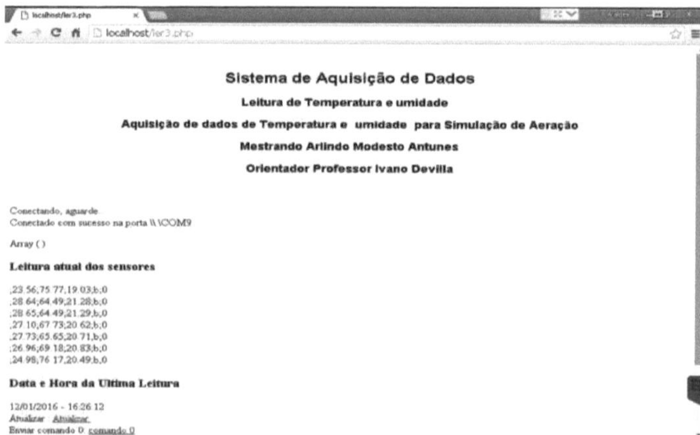

Figure 30 - Screen shot of the aeration control software used in the silo prototype. Source: author.

On the screen in Figure 31, you can see how the software is working instantly, collecting data on temperature, relative humidity and dew point temperature. Afterwards, you can see the strategy being used to aerate the stored grains (b).

Another feature developed in the system was the possibility of monitoring the "operation" or "non-operation" of the aeration fan. You can see in Figure 31 that the program shows on the screen whether the fan is running with a number (1) or not running with a number (0). This process makes it easier to make decisions and control the fan's operating hours.

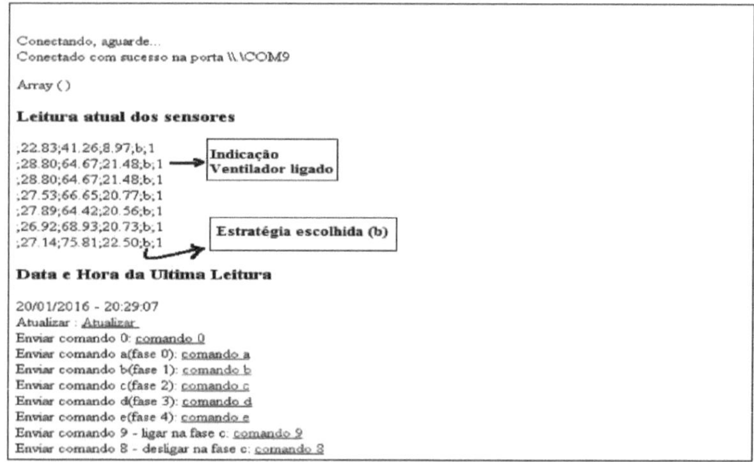

Figure 31- Aeration control software. Source: author.

The data visualization screen, Figure 32, is the main screen of the program, displaying in real time the results of the data acquisition, according to the configuration of the desired aeration strategy (phase).

As shown in Figure 32, the program has five aeration tests that can be used. These are:

a) Phase (0) - Continuous aeration;

b) Phase (1) - Room temperature control Tab < 22°C;

c) Phase(2) - Control via timer;

d) Phase (3) - Temperature and relative humidity control;

e) Phase (4) - Control via the temperature difference between the grains and the environment.

In order to provide manual control of the aeration process at any time, "command 9" was assigned to turn the fan on and "command 8" to turn it off. As a result, the process of activating any of the program's functions is done by clicking on the "command" option, as shown in Figure 32.

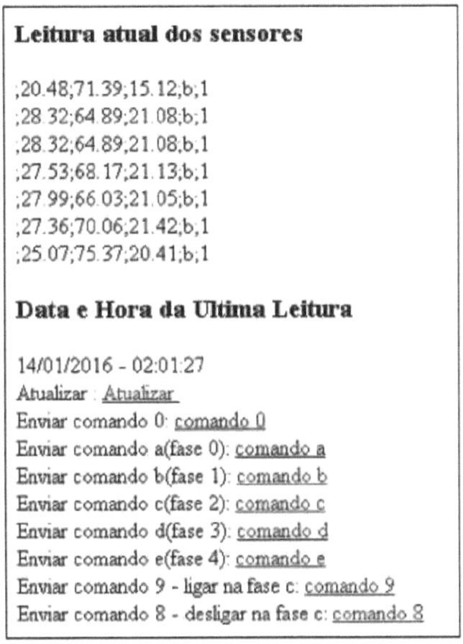

Leitura atual dos sensores

;20.48;71.39;15.12;b;1
;28.32;64.89;21.08;b;1
;28.32;64.89;21.08;b;1
;27.53;68.17;21.13;b;1
;27.99;66.03;21.05;b;1
;27.36;70.06;21.42;b;1
;25.07;75.37;20.41;b;1

Data e Hora da Ultima Leitura

14/01/2016 - 02:01:27
Atualizar : Atualizar
Enviar comando 0: comando 0
Enviar comando a(fase 0): comando a
Enviar comando b(fase 1): comando b
Enviar comando c(fase 2): comando c
Enviar comando d(fase 3): comando d
Enviar comando e(fase 4): comando e
Enviar comando 9 - ligar na fase c: comando 9
Enviar comando 8 - desligar na fase c: comando 8

Figure 32- Screen for reading data and commands for the aeration phases. Source: author

Thus, the computer program described, using the PHP programming platform, proved to be efficient in the data transmission management process and for the actions related to the control strategy of the aviation system, agreeing with authors who have worked with the same purpose, (LAWRENCE and MAIER, 2011).

Lopes et al. (2010) and Junior (2011) also corroborate the efficiency of computer programs, both of which developed a control system for aerating stored grains and concluded that they provide efficient control, cool the mass of grains and minimize oscillations in the temperature and water content of the grains.

4.6. Control strategies

During storage, environmental and storage data in the prototype silo were collected to evaluate the aeration process. It can be seen that the grain mass cooled down in all four aeration strategies, regardless of the storage period, considering that all the simulations started with grain mass temperatures in the 30 to 32 °C range. This study corroborates other authors who have also found the same trend of grain mass cooling (DEVILLA et al., 2004; LAWRENCE and MAIER, 2011).

Figure 33 shows the ambient meteorological data collected during the tests of the control strategies studied.

It can be seen that the city of Anápolis - GO, which has a tropical climate, has remarkable characteristics when it comes to temperature and relative humidity. At night, it was possible to notice a reduction in the temperature of the environment, coupled with an increase in relative humidity. This characteristic shows the need for efficient control of the aeration process of stored grains (LOPES et al., 2015).

In the control strategy, continuous aeration (a), the ambient temperature ranged from 20.3 to 39.0°C and relative humidity from 12.74 to 43.7%, a fact that differed from the other strategies, where the averages ranged from 18 to 32°C and relative humidity from 30 to 95%, which can be explained by the start of the rainy season in the region.

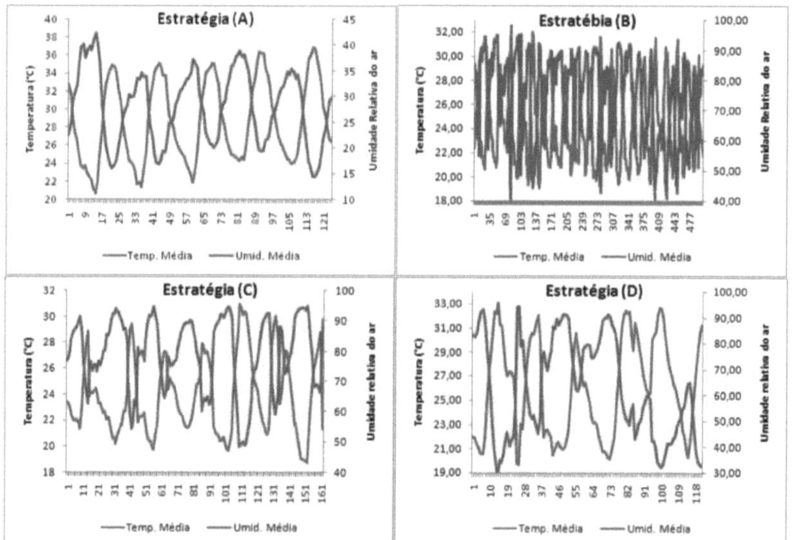

Figure 33- Average ambient temperature and relative humidity during aeration strategies a, b, c and d, in hours. Source: author.

Figures 34, 35, 36 and 37 show the evolution of grain temperature during the storage period for the control strategies evaluated.

With regard to the temperature of the grain mass, it was noted that at the beginning of the period in which the system was in operation there were sudden variations in the temperatures monitored inside the silo due to internal gradients, the difference in temperature between the inside and outside of the silo and the probable action of insects. These temperature variations in the grain mass were observed in all strategies, Figures 34, 35, 36 and 37.

In the continuous aeration process, aeration was carried out for a period of six days, reaching an average grain mass temperature of [23.3±0.54] °C. Temperatures lower than this could not be reached, as the ambient temperatures during the aeration process remained at a minimum of 20.5 °C and a maximum of 39 °C.

In the continuous aeration strategy (a) and in the ambient air temperature control (b), the cooling process of the grain mass can be seen intensely from 24 until
48 hours of aeration. Devilla et al. (2004) and Nascimento and Queiroz (2011) also verified this characteristic when studying the aeration of stored corn grains.

According to Figures 34, 35, 36 and 37, it can be seen that in all the aeration tests, the cooling direction of the mass is vertical, upwards in the silo, in agreement with Lawrence and Maier (2011) who also observed this fact when studying fifteen grain aeration strategies. Thus, it is clear that the cooling front of the grain mass starts from the base of the silo up to a height of 1.5 m, becoming slower after this point.

In Figure 36, it was possible to see that when the aeration was controlled via a timer during the night, from 10 p.m. to 10 a.m., there was greater control of the temperature gradient of the grain mass, thus extending the aeration period by nine days. This fact was also pointed out by Lopes et al. (2010) when they studied the effects of different control strategies on the aeration environment.

It was found that during the aeration process, for all simulations, the heights (1 and 2) inside the silo showed similar temperature and relative humidity during the process. It is known that the deeper parts of the silo tend to heat up very slowly, due to their low thermal conductivity, Elias (2002), Santos (2002); Quirino et al. (2013), however, it is known that the upper parts of the silo are influenced by the storage environment.

Figure 37 refers to the aeration control strategy using the temperature difference between the grains and the environment. In this figure, it can be seen that within six days the grains were cooled, which may have been due to the long periods of low temperature in the environment and the control strategy proving to be efficient (LOPES et al., 2010).

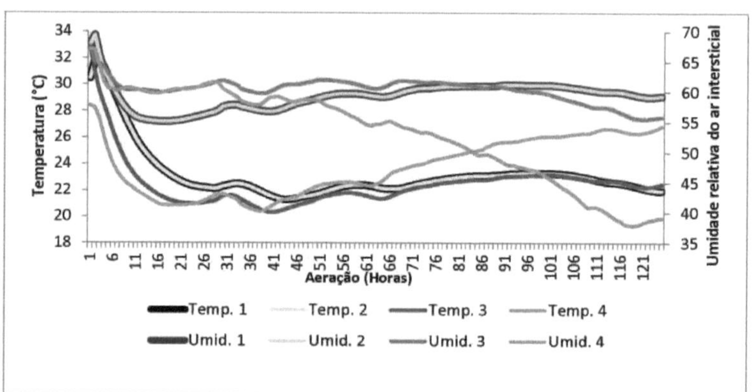

Figure 34 - Average interstitial grain air temperature and relative humidity at four silo heights (1, 2, 3, 4) during the continuous aeration control strategy. Source: author.

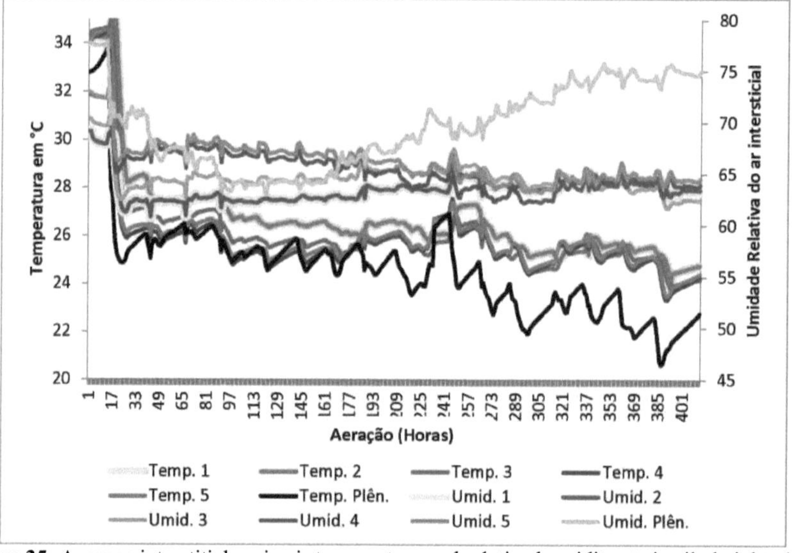

Figure 35- Average interstitial grain air temperature and relative humidity at six silo heights (1, 2, 3, 4, 5, Full), during the ambient air temperature control strategy. Source: author.

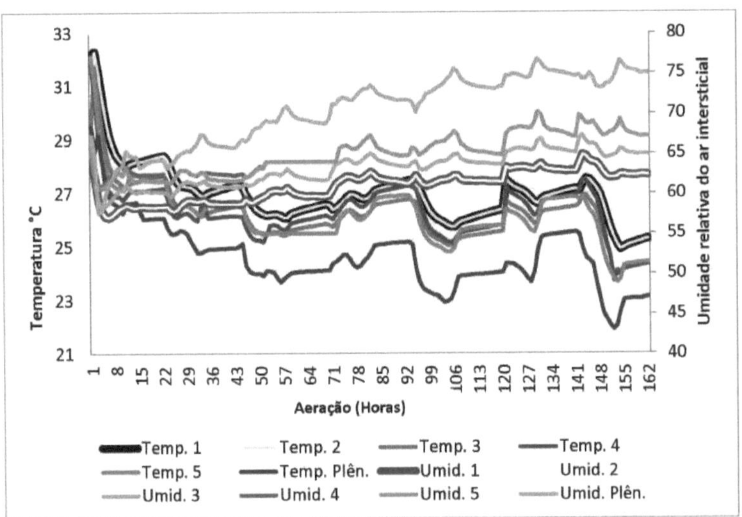

Figure 36 - Average interstitial air temperature and relative humidity of grains at six silo heights (1, 2, 3, 4, 5, Full), during night aeration control strategy. Source: author.

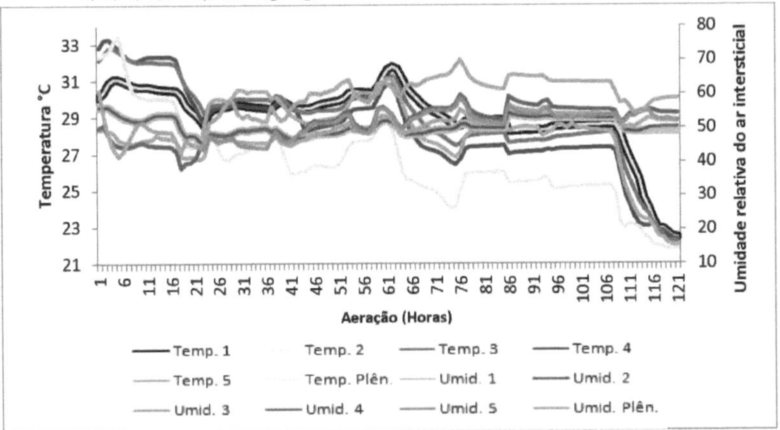

Figure 37 - Average interstitial air temperature and relative humidity of grains at six silo heights (1, 2, 3, 4, 5, Full), during temperature control strategy via temperature difference between grains and ambient air. Source: author.

Figures 38, 39, 40 and 41 show the variation in grain water content in "% b.u." at six different heights in the grain storage prototype during the test period.

Based on Figure 38, continuous aeration, it can be seen that there was overdrying in the lower profile of the grain mass, near the base of the silo, however, during the aeration process, the grains lost an average of 1.6 %b.u., ending the process with [10.39±1.2] %b.u.

It can be seen that the upper part of the silo, heights 1 and 2, are entirely exposed to the weather, gaining or losing water due to direct contact with the environment.

It was also found that aeration at night was the best way of preserving the average water content of the grains, ranging from [11.2±0.64 to 11.4±0.33] % b.u. at the end of the aeration process. This is justified by the environmental conditions (low night-time temperatures combined with relative humidity of between 50 and 65%).

In the control via temperature difference between the grains and the ambient air, the aeration process began with an average water content of [10.55±0.8] % b.u., and ended with [11.22±1.0] % b.u. On the other hand, aeration via temperature control of the external environment enabled an average variation in the water content of the grains from [12±0.51 to 10.88±0.62] % b.u.

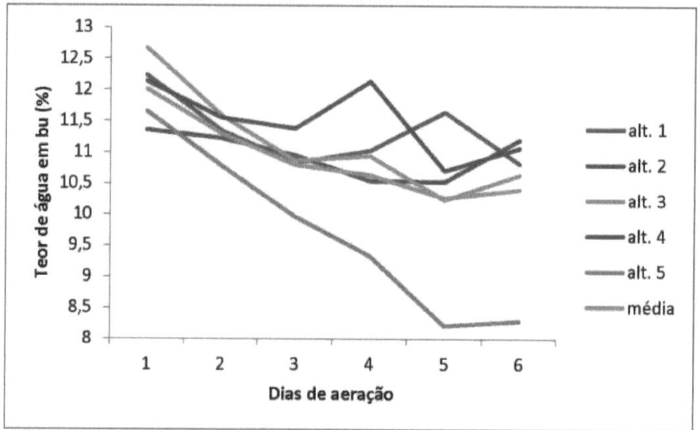

Figure 38 - Grain water content in % b.u., related to six silo heights and aeration days, during the continuous aeration control strategy. Source: author.

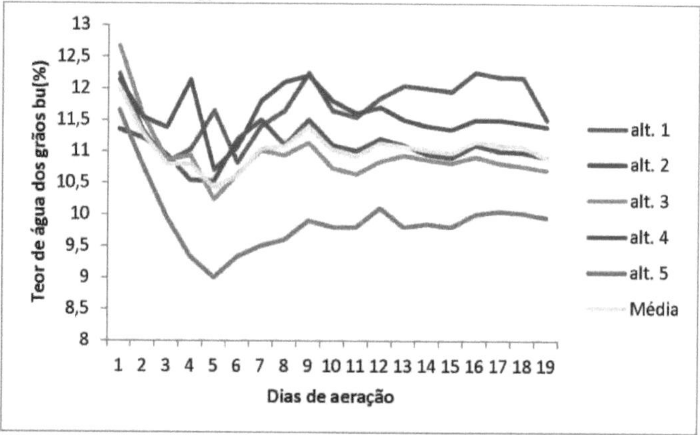

Figure 39 - Grain water content in % b.u., related to six silo heights and days of aeration, during the ambient air temperature control strategy. Source: author.

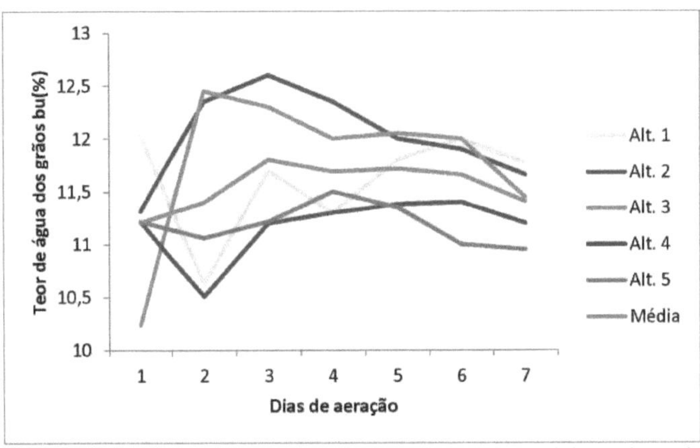

Figure 40 - Grain water content in % b.u., related to six silo heights and days of aeration, during the night aeration control strategy. Source: author.

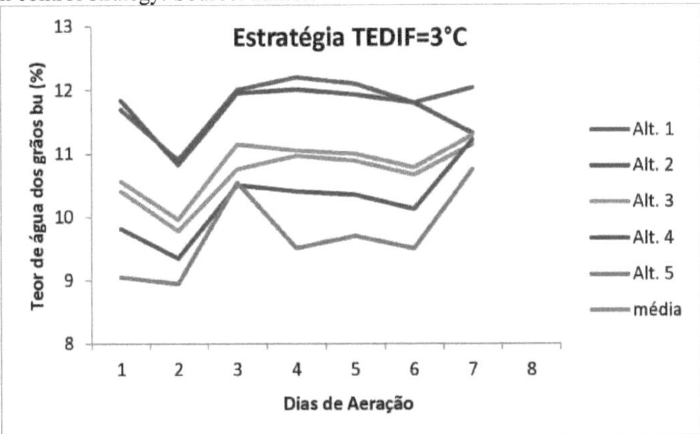

Figure 41 - Grain water content in bu (%), related to six heights in the silo and days of aeration, during the temperature control strategy via the temperature difference between the grains and the ambient air. Source: author.

In the strategies, air temperature control (b), night aeration (c) and control via the temperature difference between the grains and the environment (d), it was possible to achieve temperatures that significantly reduce the proliferation of insect pests in the stored grain mass, Table 4.

It is known that temperatures between 27 °C and 34 °C are permissible for the proliferation of most stored grain insect pests in tropical and subtropical regions (ARO, 2008). However, their growth is suppressed by temperatures between 17 °C and 22 °C (GARCIA et al., 2000; NAVARRO and NOYES 2010). It was therefore found that strategy d, temperature control between the grains and the environment < 3 degrees Celsius, was the best strategy, followed by strategy B, ambient air temperature control.

Table 4 - Averages and standard deviation of the temperature in strategies a, b, c and d related to the heights in the silo.

	Silo height						
Strategy	1	2	3	4	5	Alt. Plen.	Average
a	23,6±1,99	23,6±1,99	23,3±2,17	22,5±1,60	-	-	23,3±0,50
b	24,0±1,91	23,8±1,92	23,9±1,77	23,3±1,87	23,5±1,76	20,6±2,21	23,2±0,85
c	24,8±1,06	24,8±1,06	23,2±0,95	23,8±1,06	23,5±1,05	21,9±1,94	23,7±0,85
d	22,4±1,51	22,4±1,51	22,2±1,43	22,3±2,12	22,0±1,98	21,8±2,31	22,2±0,95

Figure 42 shows the period of operation of the aeration fan, in which aeration strategy d, controlling the aeration via the temperature difference between the grains and the ambient air, proved to be the most economical in terms of electricity costs, given that the cooling of the grain mass using strategy "d" was completed after 60 hours of ventilation. Next, the night aeration strategy was also economical, spending 73 hours to cool the grain mass. Similar results were found by Lawrence and Maier (2011), when they evaluated various controls for aerating stored grain and concluded that flow rates of 0.1 $m^{3.}$ min^{-1} t^{-1} combined with temperature control strategies result in lower energy consumption.

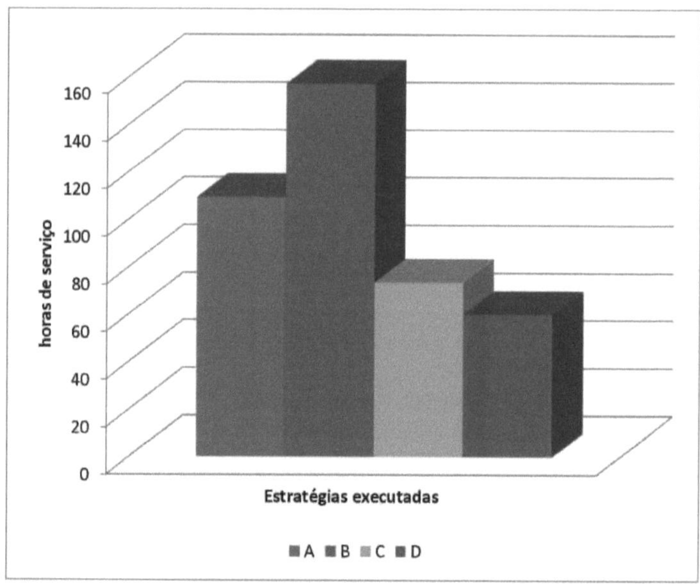

Figure 42- Number of hours with the fan on in the different aeration strategies evaluated. Source: author.

CONCLUSION

According to the results obtained and under the conditions in which this work was carried out, it can be concluded that:

1. The operational results obtained with the computer program developed proved to be efficient for controlling the aeration processes of grains stored in the prototype.

2. The prototype for storing and controlling the aeration of stored grain was efficient in the process of controlling grain storage.

3. The control strategies, aeration of grains via a temperature difference between the grains and the environment (TDIF=3°C) and aeration at night (9 to 10 p.m.), proved to be efficient for the aeration process in the Anápolis - GO region.

4. Temperature and relative humidity data relating to the storage of corn grains were collected and proved the efficiency of the computational and structural system.

LITERATURE REVIEW

AGRICULTURAL RESEARCH ORGANIZATION - ARO. Aeration and cooling for control ofstored graininsects . Available at: <http://www.agri.gov.il/Envir/aeration/aeration.html>. Accessed on: 5 Jan. 2014.

ALENCAR, E. R. Effect of storage conditions on the quality of soybeans (*Glucine max* (L.) MERRIL) and crude oil. Thesis (Doctorate). Federal University of Vinosa. Vinosa, July 28, 2006.

ARAÚJO, W. L., JÚNIOR, J. R. S., ALEIXO, D. L. & LOPES, K. P. Diagnosis of seed storage on small farms in the municipality of Pombal - PB. **Revista Verde de Agroecologia e Desenvolvimento Sustentável**, ed.3, p. 169-175. 2012.

BAAL, E. Recomendares para projeto de unidades de beneficiamento e armazenagem de graos com enfoque em segurança do trabalho. 2014. (monografia de especializagao)(engenharia de seguranza no trabalho) - Universidade Regional do Noroeste do Estado do Rio Grande do Sul, Ijuí, 2014.

BARRETO, A. A.; ABALONE, R.; GASTÓN, A. MathematicalModelling of Momentum, Heat and masstransfer in grainsstored in silos. Part II: modelApplication. **Latin American Applied Research**. Ed. 4, V.43, 2013.

Brazil. Ministry of Agriculture and Agrarian Reform. Rules for seed analysis, Brasília, DF, 399 p. 2009.

BROOKER, D. B.; BAKKER-ARKEMA, F. W.; HALL, C. W. Drying and storage of grains and oilseeds. New York: Van Nostrand Reinhold, 1992.

BROOKER, D. B.; BAKKER-ARKEMA, F. W.; HALL, C. W. Drying cereal grains. 5 ed. Westport, The AVI Publishing Company, 1982.

CASADA, M. E.; ARTHUR, F. H.; AKDOGAN, H. Temperature monitoring and aeration strategies for stored wheat in the central plains. In: ASAE ANNUAL INTERNATIONAL MEETING/CIGR WORLD CONGRESS. 9 p,. 2002.

COMPANHIA NACIONAL DE ABASTECIMENTO (Brazil). Grain survey - 2014/15 harvest. Brasília, v.1,n.11,   p.1-82.   Available at: <http://www.conab.gov.br/conteudos.php?a=1253& ... > [Accessed August 10, 2014].

CAVALCANTE, M. A.; TAVOLARO, C. R. C.; MOLISANI, E. Physics with Arduino for beginners. **Revista Brasileira de Ensino de Física**, v. 33, n. 4, p. 4503, 2011.

DELMÉE, Gérard Jean. Flow measurement manual. Sao Paulo: Edgard Blücher, 2. ed, 476 p. 1982.

DEVILLA, I. A.; COUTO, S. ZOLNIER, S.; SILVA, J. S. Variation of temperature and humidity of grains stored in silos with aeration. **Revista Brasileira de Engenharia Agrícola e Ambiental**, v. 8, n. 2-3, p. 284-291, 2004.

DILLY, Rosiene Oliveira; MENDES, Luiz Felipe Carvalho. Real-Time Application of Humidity and Temperature Monitoring Using Arduino. **Caderno de Estudos em Sistemas de Informado**, v. 2, n. 1, 2015.

ELIAS, M. C. Grain storage on small and medium scales. Pelotas: Editora da UFPEL, 218 p. 2002.

EVANS, M.; NOBLE, J.; HOCHENBAUM, J. Arduino in action. . Novatec Editora, Sao Paulo, 1 ed. 412 p. 2013.

FURQUIM, L. C.; CASTRO, C. F. d. S.; RESENDE, O.; Campos, J. M. C. Effect of drying and storage of Pinhao-manso (Jatroph curcas L.) seeds on oil quality. **Revista científica**. n.1 p.51 - 70, 2014.

GARCIA, M. J. D. M.; FERREIRA, W. A.; BIAGGIONI, A. M. Development of insects in corn stored in a sealed system. **Arquivos do Instituto Biológico**, v.67, p.63-75, 2000.

HAGSTRUM, D.W.; REED, C.; KENKEL, P. Management of stored-wheat insect pest. **Integrated Pest Management Review**. v. 4, n. 2, 127-142. 1999.

MAURO BORGES INSTITUTE - IMB (2012). Electronic annals. Available at:<http://www.seplan.go.gov.br/sepin/goias.asp?id_cad=6000>. Accessed on: September 22, 2014.

JAYAS, D. S.; WHITE, N. D. G.; MUIR, W. E. Stored-Grain Ecosystem. **Drying Technology**, v. 13, n. 4, p. 1045-1046, 1995.

JAYAS, Digvir S.; WHITE, Noel DG. Storage and drying of grain in Canada: low cost approaches. **Food control**, v. 14, n. 4, p. 255-261, 2003.

JUNIOR, J. E. K. Simulation and control of the aeration system of soybean grain mass. 205 p. Dissertation (Master's Degree) - Regional University of the Northwest of the State of Rio Grande do Sul. Ijuí, 25/02/ 2011.

KALIYAN, N.; MOREY, R.V.; WILCKE, W.F.; CARRILO, M.A.; CANNON, C.A. Low-temperature aeration to control Indianmeal moth, Plodiainterpunctella (Hübner), in stored grain in twelve locations in the United States: a simulation study. **Journal of Stored Products Research**, v.43, n. 2, p.177-192, 2007.

KNOB, A. H. Application of digital image processing to analyze grain mass anisotropy. 2010. 71 p. Dissertation (Master's degree in mathematical modeling) - Department of Physics, Statistics and Mathematics. Universidade Regional do Noroeste do Estado do Rio Grande do Sul, Ijuí.

KHATCHATOURIAN, O. A.; Binelo, M. O.; FAORO, V.; TONIAZZO, N. A. Threedimensional simulation and performance evaluation of air distribution in horizontal storage bins. **Biosystems Engineering**, v. 142, p. 42-52, 2016.

LACERDA FILHO, A. F.; AFONSO, A. D. L. Some considerations on the aeration of agricultural grains. Lecture **notes**. Federal University of Viçosa. 37 p, 1992.

LASSERAN, J. C. Grain aeration. Centreinar Series. Artes Gráficas Formato AS, Belo Horizonte , MG. n. 2, p. 1-131, 1981.

LAWRENCE, J.; MAIER, D. E. Aeration strategy simulations for wheat storage in the sub-tropical region of north India.**Transaction of the ASABE**, v. 54, p. 1395-1405, 2011.

LOPES, D. de C.; NETO, A. J. S.; JÚNIOR, R. V. Comparison of equilibrium models for grain aeration. **Journal of Stored Products Research**, v. 60, p. 11-18, 2015.

LOPES, D. C. REAL-TIME SIMULATION AND CONTROL FOR GRAIN AERATION. 2006. 135p. Thesis (Doctorate in Agricultural Engineering) - Federal University of Vigosa, Vigosa, 2006.

LOPES, D. C.; MARTINS, J. H.; MONTEIRO, P. M. B.; FILHO, A. F. L. Effect of different control strategies on the grain storage environment in tropical and subtropical regions. **Revista Ceres**, Vigosa, v. 57, n.2, p. 157-167. 2010.

LOPES, D. C.; MARTINS, J. H.; LACERDA FILHO, A. F.; MELO, E. C.; MONTEIRO, P. M. B.; QUEIROZ, D. M. Aeration strategy for controlling grain storage based on simulation and real data acquisition. **Computers and Eletronics in Agriculture**. v. 63, p.140-146. 2008.

LUIZ, M. R. Theoretical and experimental study of tomato (Lycopersiconesculentum) drying. 2012.160 p. PhD thesis in Mechanical Engineering. Federal University of Paraíba, Joao Pessoa, PB.

MACINTYRE, A. J. Industrial Ventilation and Pollution Control. Editora LTC. Sao José - SC, 2 ed, 1990.

MAGAN, N.; ALDRED, D. Post-harvest control strategies: minimizing mycotoxins in the food chain. **International Journal of Food Microbiology**, v. 119, n. 1, p. 131-139, 2007.

MOREIRA, R. G. Grain aeration using natural and cold air. In: **International Symposium on Grains Conservation. Proceedings, Canela**. p. 177-196, 1993.

MUIR, W. E.; JAYAS, D. S. Temperatures of stored grains and oilseeds.National Science Programs. Available at:

URL:http://res2.agr.ca/winnipeg/storage/pubs/presbios/chap08rf.pdf. Consultation carried out in September 2014.

NASCIMENTO, V. R. G.; QUEIROZ, M. R. Aeration strategies for stored corn: Temperature and water content. **Engenharia na Agrícultura**.v. 3, n.4. Jaboticabal. 2011

NAVARRO, S.; CALDERON, M. Aeration of grain in subtropical climates. Rome: FAO, Agricultural Services Bulletin, n. 52, 119p. 1982.

NAVARRO, S.; NOYES, R. T. The mechanics and physics of modern grain aeration management. Boca Raton, CRC Press, 2 ed, 647 p. 2010.

PUZZI, D. **Grain supply and storage**. Campinas: Updated edition. Instituto Campineiro de Ensino Agrícola, 2003. 666 p.

QUIRINO, J. R.; MELO, A. P. C.; VELOSO, V. R. S.; ALBERNAZ, K. C.; PEREIRA, J. M. Artificial cooling in the conservation of the commercial quality of stored corn grains. **Revista Engenharia Agrícola.** Bragantia, v.72, n.4, p. 378-386, 2013.

RANALLI, R.P.; HOWELL JÚNIOR, A.; ARTHUR, F.H.; GARDISSER, D.R. Controlledambientaerationduringricestorage for temperature and insectcontrol. AppliedEngineering in Agriculture, Michigan, v.18, n. 4, p.485-490, 2002.

REED, C.; ARTHUR, F. H. AERATION. In: Subramanyam, B.; Hagstrum, D.W. Alternative to pesticides in stored-product IPM. Norwell: Kluwer Academic Publishers, p.51-72, 2000.

RIGO, A. D.; RESENDE, O.; OLIVEIRA, D. E. C.; DEVILLA, I. A. Control Strategies for the Grain Aeration Process in Metal Silos Prototypes. **Gl. Sci. Technol**, Rio Verde, v. 05, n. 03, p. 47-55, Sep/Dec. 2012.

SAUER, D.B. **Storage of cereal grains and their products**. 1992. 615p

SANTOS, D. L.; CENTENARO, M. Analysis of the soybean storage system: in-house vs. outsourced. ENCONTRO DE INICIAÇÂO CIENTÍFICA-ENIC, n. 6. 2014, Dourados, Anais... Dourados, 2014.

SANTOS, J. P. Preventive methods for controlling stored grain pests. In: LORINE, I.; MIIKE, L. H.; SCUSSEL,V.M. (Ed.). Grain storage. Campinas: Instituto Bio Geneziz, 2002. chap. 3.3, p.157-174.

SILVA, J.S.; LACERDA FILHO, A.F.; DEVILLA, I. A. Aeration of stored grains. In: Silva, J. S. **Secagem e armazenagem de produtos agrícolas**. Viçosa: Editora Aprenda Fácil, p.261-277. 2000.

SILVA, J.S.; LACERDA FILHO, A.F.; DEVILLA, I.A.; LOPES, D.C. Aeration of stored grains. In: SILVA, J.S. **Secagem e armazenagem de produtos agrícolas.** Viçosa: Editora Aprenda Fácil, p.269-295, 2008.

USATIKOV, S. B.; SHAZZO, A. Y.; TIVKOV, M. A. Mathematical modeling of self-heating of grain mass. **Kuban State University of Technology**. n. 4. 2003. Available at:< http://cyberleninka.ru/article/n/matematicheskoe-modelirovanie- protsessov-aktivizatsii-sploshnogo-samosogrevaniya-zernovoy-massy>. Accessed on: 20/10/2015.

VASCONCELLOS, M. B. Mathematical modeling of air flow in particulate media under homogeneous and anisotropic conditions. 2012. 76p. Master's dissertation (mathematical modeling). Regional University of the Northwest of the State of Rio Grande do Sul. Ijuí, RS.

WALIA, E. S.; GILL, E. S. K. A Framework for Web Based Student Record Management System using PHP. International Journal of Computer Science and Mobile Computing. IJCSMC, Vol. 3, N. 8, pg. 24-33, 2014.

WEBER, E. T. Excelencia em beneficiamento e armazenagem de graos. Canoas: Salles, 2005.

WHITE, C. **Wireless Grain Silo Monitoring and Control**. 2013, 50p. Final course work (Electrical and Electronic Engineering) - University of Southern Queensland, 2013.

ZHANGXIAO RU. Bulk rice quality and development of a quantity monitoring system. 2013,182p. Dissertation (Master's Degree in Engineering Electrical) - National Taiwan University, 2013.

ZULKIFLI, N. S. A.; HARUN, FK Che; AZAHAR, N. S. XBee wireless sensor networks for Heart Rate Monitoring in sport training. In: Biomedical Engineering (ICoBE), 2012 International Conference on. IEEE, 2012. p. 441-444.

APPENDIX

To control the fan, it was necessary to attach a contactor and thermal relay for the aeration system to function fully.

Figure 43 - Contactor 3RT1015 and relay 3RU1116 wired for the project. Source: author.

I want morebooks!

Buy your books fast and straightforward online - at one of world's fastest growing online book stores! Environmentally sound due to Print-on-Demand technologies.

Buy your books online at
www.morebooks.shop

Kaufen Sie Ihre Bücher schnell und unkompliziert online – auf einer der am schnellsten wachsenden Buchhandelsplattformen weltweit! Dank Print-On-Demand umwelt- und ressourcenschonend produziert.

Bücher schneller online kaufen
www.morebooks.shop

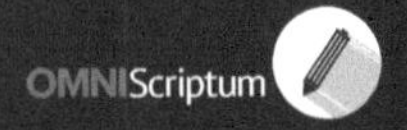

Printed by Books on Demand GmbH, Norderstedt / Germany